All About Sound

Copyright © by Harcourt, Inc.

All rights reserved. No part of this publication may be reproduced or transmitted in any form or by any means, electronic or mechanical, including photocopy, recording, or any information storage and retrieval system, without permission in writing from the publisher.

Requests for permission to make copies of any part of the work should be addressed to School Permissions and Copyrights, Harcourt, Inc., 6277 Sea Harbor Drive, Orlando, Florida 32887-6777. Fax: 407-345-2418.

HARCOURT and the Harcourt Logo are trademarks of Harcourt, Inc., registered in the United States of America and/or other jurisdictions.

Printed in Mexico

ISBN 978-0-15-362234-2

ISBN 0-15-362234-2

6 7 8 9 10   0908   16 15 14 13 12
4500356340

SCHOOL PUBLISHERS

Visit *The Learning Site!*
www.harcourtschool.com

Good Vibrations

Take a rubber band. Stretch it between your fingers. Then pluck it. It makes a sound you can hear.

The part of the rubber band you plucked moves back and forth. It moves so fast that all you see is a blur. This back-and-forth movement is called a **vibration**.

All sounds are produced by vibrations. Plucking the rubber band made it vibrate, and you heard the sound it made. The sounds from guitars and drums are produced by vibrations. A saxophone being played, a dog barking, and even a book dropping on the floor—all of these sounds are produced by vibrations. Loud or soft, near or far, high or low—all sound works this way.

The sound from the rubber band traveled through the air to your ear so that you could hear it. Sound travels through other materials, too.

If someone stands on one side of a closed door and talks, you might be able to hear the person. The sound this person makes travels through the door, a solid.

If you swim underwater, you might hear sounds, too. The sounds you hear travel through water, a liquid.

The vibrations that produce sound can travel through solids, liquids, and gases, such as air.

 MAIN IDEA AND DETAILS What happens to a rubber band when you pluck it?

When you pluck a guitar string, the string vibrates and makes sound.

Catch the Wave

Sound waves move outward from the vibrating object. Imagine what happens to a pond when you throw in a rock. The rock hits the water and pushes the water particles. Those particles push on neighboring water particles, and the water ripples. The movement of the water particles produces rings of ripples, moving through the water.

Where the rock hits the water, it transfers energy into the water. At that point, the waves are close together and move fast. As the waves move out across the pond, they spread farther apart.

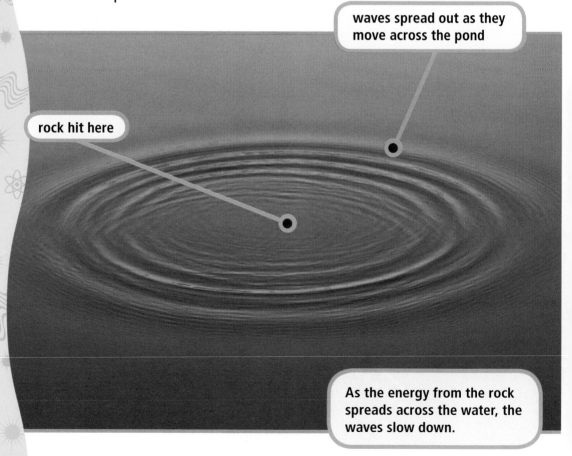

waves spread out as they move across the pond

rock hit here

As the energy from the rock spreads across the water, the waves slow down.

Sound waves reach every part of this concert hall when music is played.

This same process can happen with sound waves. The sound waves made by a vibrating object spread out through the air and through many of the materials they encounter. Eventually, they reach your ear and then your brain.

 COMPARE AND CONTRAST How are sound waves like the ripples in a pond?

Fast Fact

Designers of concert halls must consider how sound waves will behave in every part of the room. Before the actual hall is built, miniature models are made to test the quality and direction of the sound waves.

Looking at Sound

Scientists who study sound spend a lot of time looking at it. They use different kinds of scopes and machines to watch sound waves. They can look at the shape and movement of a wave and tell what kind of sound that wave describes.

You can start by measuring a sound's wavelength. In a sound wave, the **wavelength** is the distance between one point on one wave and the identical point on the next. A water wave looks different but behaves the same as a sound wave. To picture a wavelength, it's helpful to use water waves as a model.

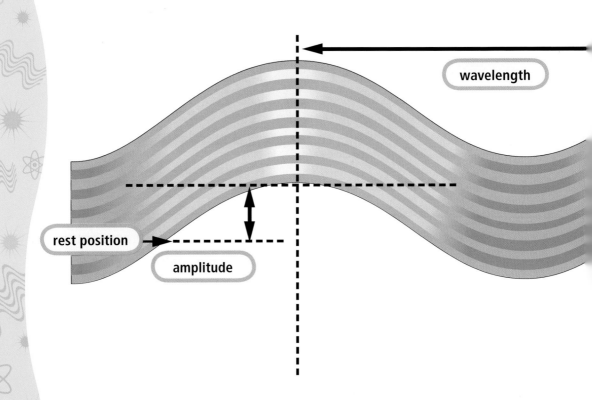

Frequency means the number of waves that pass in a second. If you counted the number of water waves that pass a particular place in a second, you would be counting frequency.

The sound wave's amplitude affects the intensity of a sound. **Amplitude** measures the amount of energy in a wave. Sound waves with larger amplitudes make louder sounds.

 COMPARE AND CONTRAST What is the difference between a high-frequency sound wave and a low-frequency sound wave?

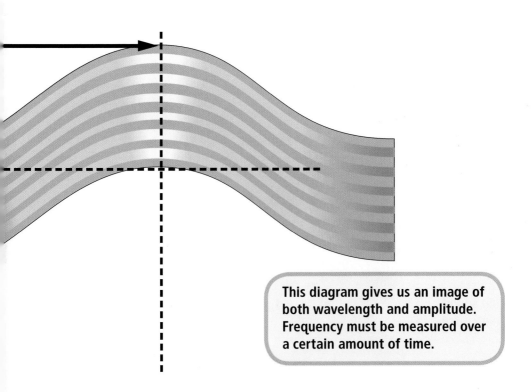

This diagram gives us an image of both wavelength and amplitude. Frequency must be measured over a certain amount of time.

Pitch and Intensity

What's the difference between a lion's roar and a kitten's meow? The lion makes a sound that is low, and the kitten makes a sound that is high.

When sound waves are farther apart, we say that the sound has a low pitch. When sound waves are close together, we say that the sound has a high pitch. **Pitch** tells how high or low a sound is.

 COMPARE AND CONTRAST What is the difference between a high-pitched sound wave and a low-pitched sound wave?

The tiny vocal cords in this kitten's throat vibrate quickly, producing sound with a high pitch.

What's the difference between a whisper and a yell? One sound is soft and one sound is loud. The **intensity** of a sound measures how loud it is. Sound waves have energy, and the amount of energy determines the intensity of the sound they will produce.

The more energy in a sound wave, the higher its intensity and therefore the louder the sound.

> **Fast Fact**
>
> Loud or soft, sound travels through the air at the same rate—1.6 kilometers (1 mi) in 5 seconds.

COMPARE AND CONTRAST What is the difference between pitch and intensity?

The lion's large vocal cords produce sound with a low pitch.

Changing Sound

Just by changing the size or tension (tightness) of the vibrating material, you can change the pitch of a sound. A musician may tune a guitar string by tightening it. Tightening the string increases the tension on it. When the tension increases, the guitar string vibrates more quickly. This increases the frequency of the sound wave and raises the pitch.

If you've ever watched a violinist play, remember how the violinist's fingers pressed down on different strings and in different places. Pressing on the string changes the length of the vibrating string. These different lengths make different vibrations— and different pitches, as well.

Loosening or tightening each string on the cello will change its pitch.

The thinner strings make a higher-pitched sound than the thicker strings.

A musician can also get a higher sound from a thinner guitar string, and a lower sound from a thicker guitar string. That's because a thin string will vibrate faster than a thick string.

 CAUSE AND EFFECT How does shortening the length of a string on a guitar change the sound the string makes?

Can You Hear Me?

You're standing outside. You shut your eyes to listen to the world around you. You hear a car starting, a bird singing, someone talking on a cell phone. You recognize all of these sounds without having to see what is making them. But even before you know what you're listening to, an amazing process called transmission has taken place.

Transmission is the process by which sound waves move through material. For example, the sound waves that you hear on the street were made when each object started to vibrate. They were transmitted through the air, all the way to your ear. Your ear continued the process all the way to your brain.

> **Fast Fact**
>
> Dogs can hear sounds with a far greater range of frequencies than humans can. They can hear any sound that has a frequency of between 15 and 50,000 vibrations per second. Humans can hear only sounds that have frequencies between 80 and 1,500 vibrations per second.

 MAIN IDEA AND DETAILS What is transmission?

The speaker in this boom box can transmit sounds a long distance.

When you hear sound, your outer ear catches sound waves and funnels them inside to your eardrum.

The sound waves cause your eardrum to vibrate. The vibrations travel along three tiny bones, called the hammer, the anvil, and the stirrup.

The three bones amplify the vibrations from your eardrum into larger vibrations that travel to the cochlea in your inner ear. Special cells in the cochlea change the vibrations into nerve signals. These signals then travel along a nerve to your brain, where they are recognized as sound.

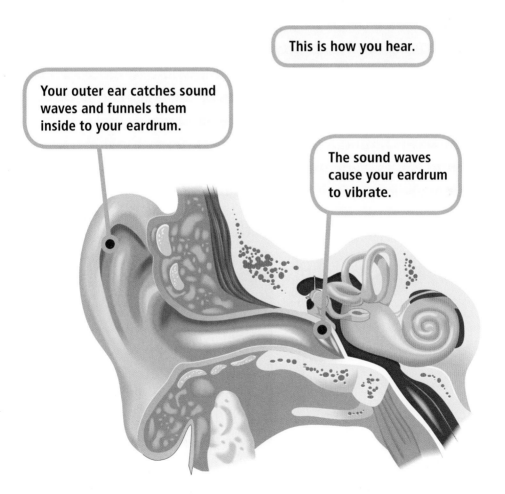

This is how you hear.

Your outer ear catches sound waves and funnels them inside to your eardrum.

The sound waves cause your eardrum to vibrate.

Sounds and Surfaces

Any material that vibrates can transmit sound. You hear someone walking in the apartment upstairs because sound is transmitted through the solid floor. Swimming under water, you hear people talking by the pool because sound is transmitted through the water.

Sound waves can behave like light waves. They can bounce. When light waves hit a mirror, they bounce back in the same pattern, and you see an exact duplicate of the pattern—a mirror image. When sound waves hit a smooth, flat surface, they bounce back in the same pattern, and you may hear an echo. This process is called **reflection**. You might hear sound reflection in the gym, where sound waves bounce off the hard, smooth surfaces of the walls and floor.

There's another kind of reflection sound waves make. Just as light waves hitting a rough surface scatter everywhere, so do sound waves. When sound waves hit a rough wall, they go off in all directions.

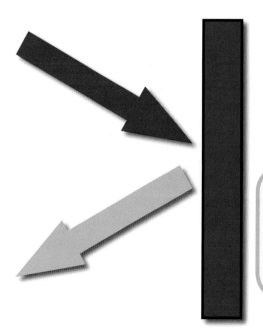

When sound waves hit a smooth, flat surface, they bounce off. When most of the sound waves go in the same direction, they make a clear echo.

Sound can also do the opposite. When sound waves travel into a carpeted, curtained room, they can seem to fade away and die. This happens because soft articles like carpets and curtains stop the sound from being reflected back into the room. This process is called **absorption**.

Sound waves behave differently in different situations. It is easier for them to pass through some materials than others. For example, sound travels better through wood than through air. In space, where there is no air and, therefore, nothing for the sound waves to cause to vibrate, there is no sound.

 COMPARE AND CONTRAST Compare sound in a bare room with hard floors and walls to sound in a room filled with carpets and curtains.

Sound will be absorbed by the carpet and curtains in this room.

Summary

Sound starts with a vibrating object. Like light waves, sound waves can be measured in terms of frequency and amplitude. Sound has pitch and intensity. When sound waves are transmitted to your ear, they travel to your brain and are recognized as sound. Sound can bounce, scatter, or echo, and sound can also be absorbed, so that it fades away or dies.

Glossary

absorption (ab•ZAWRP•shuhn) The taking in of light or sound energy by an object (15)

amplitude (AM•pluh•tood) A measure of the amount of energy in a wave (7, 15)

frequency (FREE•kwuhn•see) A measure of the number of waves that pass a fixed point in a second (7, 10, 15)

intensity (in•TEN•suh•tee) A measure of how loud or soft a sound is (7, 9, 15)

pitch (PICH) A measure of how high or low a sound is (8, 9, 10, 15)

reflection (rih•FLEK•shuhn) The bouncing of light, sound, or heat off an object (14, 15)

transmission (tranz•MISH•uhn) The passing of light or sound waves through a material (12)

vibration (vy•BRAY•shuhn) A quick back-and-forth motion (2, 3, 4, 5, 8, 10, 11, 12, 13, 14, 15)

wavelength (WAYV•length) The distance between a point on one wave and the identical point on the next wave (6)